Blockchain:

Ultimate Beginner's Guide to Blockchain Technology

Cryptocurrency, Smart Contracts, Distributed Ledger, Fintech and Decentralized Applications

Matthew Connor

Table of Contents

Introduction

Congratulations on downloading **Blockchain: Ultimate Beginner's Guide to Blockchain Technology – Cryptocurrency, Smart Contracts, Distributed Ledger, Fintech and Decentralized Applications** and thank you for doing so. While you might not yet be familiar with blockchain technology, you have almost certainly heard of the investment craze that it is responsible for. Without blockchain technology, there would be no Bitcoin or any of the other thousand cryptocurrencies that are currently driving investors of all stripes wild. Blockchain's potential extends far beyond just investing in new cryptocurrencies, however, which is why you need this book if you hope to learn enough about the technology in question to put it to work for you.

Inside you will find everything you need to know about the basics of blockchain technology to determine the best way to move forward for you or your company to put yourself in place to take advantage of the technology that is already being called the most significant technological advancement since the creation of the internet. It doesn't matter if you are interested in starting a new business or looking for a means of reinvigorating one that is beginning the flag, blockchain technology has something for you.

Blockchain technology is going to offer up new ways of looking at old problems for transferring money between parties, keeping track of files, automating contracts, counting votes, the list goes on and on. What it does at its heart, however, is using its unique blend of security and transparency to offer up a way for people to trust one another online without the help of a third party. While the exact results of what this unprecedented level of trust is going to be are not yet clear, what is clear is that for now at least, people at all levels of the supply chain can't get enough of it. This also means profit is in the air for those with the nose for it, and if you can't smell it yet, then you need to read on to ensure you don't sit on this opportunity any longer.

Chapter 1: What is Blockchain?

Now, at the beginning of 2018, it is highly unlikely that you haven't heard the term blockchain thrown around from time to time. This was likely in connection to Bitcoin and its year-end 2017 run that saw a peak price of about $20,000. This in no way means you should understand what blockchain technology is or be able to describe it in detail; and indeed, less than half the population can accurately do just that. Likewise, the overall percentage of people that interact with blockchain technology on a regular basis remains at less than half that number. Nevertheless, the following chapters will reveal why it is already called one of the most important new technologies of the twenty-first century.

First things first, a blockchain is a sort of decentralized database that works as a financial ledger. Due to its unique construction, it is incredibly secure while at the same time allowing for a virtually unlimited number of users to interact with it over time. As you may have guessed from the name, a blockchain is made up of individual blocks that each contains unique transaction information, as well as additional information relating to their specific location within the chain. A useful analogy is that a blockchain can be thought of as A Lego in that each blockchain is made up of many different parts, assembled in different ways while still falling under the same general construct.

When a transaction takes place on a blockchain, it is verified by a third party before being placed in a block with other transactions that took place in the range of a single node. When the new block is full, it is then verified by an independent third party in a process known as mining before it finds its way into the blockchain, its information is then shared between all the various nodes that are running the blockchain at a given time. If at least 51 percent of all the nodes don't agree with the new information, it is then not allowed into the blockchain proper.

All of this happens autonomously based on the code the blockchain is constructed from. Nodes transfer their data between one another using an ease of use model which means that it automatically moves from one node to the next closest node, and so on and so forth until the new block has been added to all of the currently active nodes. From a security standpoint, the way the blockchain is constructed means that to create fraudulent blocks, the fraudsters would need to create enough nodes that they constituted more than 51 percent of the whole. The cost required for such an undertaking means that it would always exceed the reward, which, in turn, means that it is unlikely that anyone will be using this means to hack blockchains anytime soon.

What's more, when a block is added to the blockchain, the block itself, along with all of its transactions, are given timestamps which means it is easy for users to determine precisely when a particular transaction takes place. Altogether, this process means that each blockchain can operate in a completely autonomous fashion, with transactions being passed to blocks, being passed to the blockchain, and being policed for errors without any guiding hand having a say in any part of the process.

Chapter 2: How Does the Blockchain Technology Work – In Plain Words

If you are still confused by the core concept of what a blockchain is exactly, try this on for size:

A person (the node) has a spreadsheet full of their financial transactions (the blockchain). This spreadsheet is shared among several people (miners) which makes it distributed. When the person goes to add a new transaction to the spreadsheet, the other people are alerted to the fact, so they can double check everything is on the level before it is added to the spreadsheet.

The first person to verify that the funds exist to pay for the transaction are paid for their time (mining rewards). The verification that the transaction is legitimate (the proof of work) is sent along to show everyone else that things are as they should be. Assuming 51 percent of the people agree with the verification, the transaction is added to the spreadsheet.

If you are still slightly confused, don't worry, several of the concepts involved are somewhat ephemeral, especially if you haven't been a part of the process yourself. As such, it may be more beneficial to discuss how the technology works as opposed to discussing what it does. With this in mind, the better question then becomes, what are the benefits of using blockchain technology:

Benefits

Improves trust: Trust is at the crux of the blockchain issue, it makes cryptocurrency possible because it allows users to trust one another, despite sharing little to no information about themselves for each transaction they complete. Likewise, this extreme level of trust makes it easy for businesses to use the blockchain to complete transactions that previously required countless different layers of paperwork.

While online transactions of various types have been able to do this sort of thing for years, and people have managed just fine in the centuries before the internet, the crucial thing that sets blockchain technology apart is that it is able to generate this level of certainty without the use of any middle man. In a non-blockchain scenario, this trust is managed by a bank or other type of financial institution who would naturally have a vested interest in ensuring that things took place as anticipated, and money was securely moved between accounts. With a blockchain in place, this same level of security can be obtained, not only more quickly, but cheaper as well.

Speed: The fact that the entire transaction happens within one blockchain, as opposed to one bank contacting another bank for a transfer of funds, means that the type of transaction that can often take a day or more to clear through a bank, can be processed and verified within minutes, if not less, through a blockchain. There are going to be times when waiting for this level of verification makes sense, for example, you probably wouldn't mind waiting for your lawyer to determine that the contract you are going to sign to purchase a house is legitimate. Most of the time the added speed is going to be appreciated, however, and for a wide variety of transactions, it is just going to make more sense. To understand just what is occurring during the verification process, consider how things currently proceed when purchasing a piece of jewelry from a private seller.

Under the current system, the negotiation for the transaction would take place; the interested parties would agree on an independent verification service, send the piece of jewelry off to be verified and then wait for the results to return. Even then, the party purchasing the jewelry would have to trust that the results were not tampered with in some way, shape or form.

When dealing with a blockchain transaction, however, the blockchain automatically takes care of all of these steps and allows both parties to view each one of them as well.

Market gaps: While high-dollar purchases are typically tracked in one type of register or another, deeds on property or titles on vehicles, for example, there are thousands of smaller, though still important, transactions that have no natural middleman to keep track of them. As blockchain technology becomes more commonplace, it will be well suited to serve as the register in these circumstances, easily tracking a wide variety of items from behind the scenes.

The barrier to creating these types of registers without a blockchain backing is sure to be substantial, starting with the difficulty of determining who would hold the register, how the information would be obtained, and how it would be updated on a regular basis. From there, it would also raise numerous questions regarding how the information would remain secure and tamper-proof, while also allowing access to those who have a legitimate need to do so. As you can see, creating a new type of middleman service isn't something that could happen overnight, which is why there is so much trust placed in governments and banks, they succeeded where so many others failed.

Moving forward, however, blockchain technology will supplant these sorts of notions, both because it will make creating new registries much easier, and also because its reputation of security will remove any worries that a new registry is inferior to an existing one. When a new registry is created, the blockchain automatically distributes it every person who decides to operate a node in the new system. Each of these people will then receive the same copy of the registry so they can ultimately add new blocks to the blockchain. Even if one of these people proves untrustworthy, the way the blockchain is constructed means all the other trustworthy people would overrule the inaccurate information, protecting the quality of the blockchain as a whole.

As such, the question of trust is removed from the equation altogether, those who complete a transaction through a blockchain can know, unequivocally what is happening with their transaction 100 percent of the time. Then, once it has been accepted into the blockchain, the people will be able to rest easy knowing that as long as at least half of the copies of the blockchain out there remain uncorrupted, there is no way for it to be changed. While trusting a single person is difficult, trusting the consensus that a large number of people have already agreed on is much more palatable.

Chapter 3: How Can Blockchain Technology Be Used

Tracking Supply Chain History

For companies that deal in the large-scale logistical nightmare of moving freight back and forth, the widespread adoption of blockchain technology, specifically in an internet of things capacity, will simplify the process dramatically. This is mainly going to be the case when it comes to determining the exact location of goods while they are in route when real-time updates are difficult to come by. Not having ready access to this information, can lead to a wide variety of different delays which can serve to affect the business as a whole negatively and, at best, often results in additional costs and an undue strain on customers and vendors.

This will ultimately be solved by shipping containers that are connected to a specific blockchain that will be equipped to notify both the buyer and the seller when the product has reached various shipping milestones and will even generate digital signatures, where required. This will also serve to improve proof of provenance issues that tend to occur when a product has multiple manufacturers. Tracking various components from start to finish is virtually impossible using traditional methods, but with a blockchain system in place, it would become exceptionally easy to ensure that products are procured both legally and humanely. As the infrastructure related to the internet of things improves, even more, it will ultimately become possible to track virtually every step in the creation of every product. Someday soon, customers will be able to track products from the factory straight to their doorsteps.

Business Data Storage

Another area that blockchain technology will soon help to streamline is business storage. While companies like Storj (discussed in chapter 7) are offering solutions in this realm for individuals, studies show that companies of all sizes likely have an extensive amount of storage that is going either unused or under-utilized. Blockchain technology will allow businesses to fully utilize their existing resources without having to resort to external organizations to store potentially sensitive data while decreasing costs at the same time. This is not to say that storage is the only reason to utilize a company blockchain as if this is the only goal your company has for blockchain technology then these goals can likely be met with just a traditional cloud-based storage solution at upwards of one-tenth the cost.

Autonomous Office Networks

Another essential thing to understand about the office of the future is that it is likely going to be incredibly full of sensors. In fact, every day more and more offices are being connected to digital assistants of one type or another, and this trend towards all-around connectivity is only going to continue to grow. As with other kinds of logistical technology, the growth of the internet of things means that offices are just going to become more interconnected as time goes on.

Once these types of autonomous office networks go online, a wide swath of the equipment in an average office will be able to communicate with one another to do things like ensure that energy usage is at its most efficient, schedule upgrades, and even schedule and pay for required maintenance and supplies before the problem even becomes visible to the user.

This will save time as well as money, as many functions will be able to be automatically scheduled to take place during off-usage hours and all of the equipment will be equipped with the ability to monitor energy usage to ensure that only the correct amount of energy is used for any particular job. This will also go a long way towards preventing both the buildup of waiting for supplies to arrive and also the time lost due to crippled productivity that comes with waiting for key supplies to arrive. While this level of automation may seem like a novelty at first, it will almost always lead to a decrease in expense in the long-term.

Smart Contracts

While it may be hard to determine precisely all of the ways that smart contracts are going to benefit small businesses in the future, it is already helping major corporations in genuine ways. This means that, regardless of the size of your business, if you are looking to put blockchain to use for you, the first thing you will need to do is ensure that you don't rush into things without thinking the entire process through thoroughly. Starting out with your mind made up could easily cause you to invest in a system that you don't have the infrastructure to support fully.

When looking into opportunities related to blockchain technology, it is important to keep in mind that nothing is yet set in stone. This, in turn, means that going out of your way to investing heavily in any system that is too exclusive could be a mistake, especially if that company goes out of business. Don't accidentally buy into the Betamax of blockchain technology, take a measured approach for now, at least until things start to settle down.

Thanks to blockchain, new companies are going to have even more ways to fund their startup dreams. 4G Capital is a company that is already using blockchain as a means of facilitating small business loans to business owners in Africa. Through the Ethereum blockchain, donors can connect to recipients directly and provide them with the funds they need via a proprietary transaction system. Furthermore, in addition to providing loans for 100 percent of the asking price, in an unsecured fashion, it gives those who would otherwise not be able to get loans access to funds that are life-changing. While currently operating in a limited manner, as this idea catches on it has the potential to change a vast swath of the financial sector.

Chapter 4: Pros and Cons of Blockchain Technology

While it can be easy to get caught up in all the hype surrounding blockchain technology, it is important to take the time to seriously consider both the pros and cons before you pull the trigger.

Read data: When it comes to most centralized databases, knowing who accesses what can often be buried somewhere inside the depths of a log file that stores everything that goes on in the database. If you are looking for a way to streamline this process, depending on its settings, a blockchain can be read without actually accessing it from a database node. Once changes are made to a given block or chain, those changes are then stored in a traditional log file. If you have data that need to be secured but also need them to be regularly seen by those who can look but not touch, then a blockchain is a preferable solution.

Data alternation: When it comes to accessing a traditional database, usually all that is required is a username and password, along with more secure alternatives if the information needs it. Blockchains can be secured in the same way, along with a digital signature that is included each time a change is made, at the block level, for each new transaction the user adds, making it very easy to determine where a transaction came from and who initiated it.

Essentially, what this means is that each time a transaction is completed, the instigating user is required to digitally sign and confirm each of their transactions, though this step can be skipped if they add information to the chain from the node directly. Even if some secondary verification is not required, the IP address of every user is automatically logged and recorded for later use.

However, depending on the frequency with which data is going to need to be altered, and the number of nodes that are going to be in play, then a blockchain may not be the best choice in all cases. In a more traditional database, users with the right level of clearance will be able to rewrite or alter any data they see fit, with a log keeping track of the changes that have been made. This is not the case with a blockchain, however, as the only way to change information once it has been added to the blockchain is to simultaneously change it across more than 50 percent of all the existing nodes at the same time. If the data is not changed in this fashion, then the change will be rejected and the blockchain will self-correct to the previous version of the information.

Data backup: While traditional databases require user authorization and potentially significant amounts of additional space for backups as well, a blockchain database is automatically updated each time a new block is added to the blockchain, and that change is then reflected across all of the relevant nodes as well. As such, this makes them a natural choice for extremely important information as all of your nodes would need to go down at once for you to permanently lose any information. Depending on how much is currently being spent on data backup solutions, this type of distributed database option may ultimately end up being a less costly alternative as well as long as the amount of data that is being stored isn't going to make adding it to the chain prohibitively expensive.

Decentralization reality: One often overlooked aspect of blockchain technology is that the greater the distance between nodes, the greater the amount of time that will be required for information to be shared between the nodes specifically. As such, if multiple nodes are going to need to be contacted on a regular basis, or if information is legally required to stay within a specific geographic location, then a blockchain database might not be the most effective means of storage possible. While this won't necessarily prohibit a blockchain database from being used in these situations, it does mean that the specific location of each node is going to matter far more than the location of individual servers in a more traditional storage scenario.

This issue is only made more complicated because blockchains can also be connected to other blockchains for specific periods of time. If this occurs, then the details in both blockchains will be visible to anyone with access to either blockchain. This means that additional precautions will need to be made before giving access to anyone new to ensure you aren't opening the door far wider than you initially anticipate.

Volume of data: Regardless of the type of blockchain you decide on, the amount of data that can be stored per block is nowhere near what can be stored within a more traditional type of centralized database. Additionally, as each new block is going to need to be spread amongst all of the currently active nodes you are going to want to make an effort to keep them fairly small to prevent the verification times between nodes from becoming unwieldy. If you are looking to work with larger files, or to store a significant amount of data, then a traditional database will allow you to do so much more efficiently.

Validation: A majority of the costs associated with running a blockchain result from the costs associated with validating the transactions that appear on it. If your blockchain is going to be used internally, then the validation costs are likely going to be quite low and will be something that members of the team can help out with to mitigate the costs to just what is being paid to in electricity to validate the transactions. If there are also external transactions coming in, and the blockchain is still private, then costs can add up over time as the proof of work required becomes more and more complicated.

The next step

After looking at the specifics, assuming you decide that taking advantage of blockchain technology makes sense for you, it is important to consider exactly what you plan on using the technology for. If you are an existing business owner who hopes to get in ahead of the curve on the next big thing, then you will want to focus on the many potential ways blockchain and smart contract technology can work together to improve many of the ancillary aspects of your business. Specifically, you are going to want to take a long hard look at things that have the potential to decrease costs or improve efficiency.

This means you will also need to consider all the many ways that utilizing a blockchain will make you more competitive in the eyes of the competition by allowing you to get the jump on the emerging trends in your industry. Alternately, you will want to consider the various disruptions to the way your business works that blockchain might bring to light, and work to move things around now, so the disruption is as minimal as possible. Remember, being aware of what is likely coming next will make it easier to face head-on.

If you are looking to form a new business based on blockchain technology, then the best way to get started is to work with as many different blockchains as possible. This will not only help to improve your grasp of the technology but will also help to make the technology more mainstream, which is what is needed for new blockchain companies to really take off. If you ever hope to break into the mainstream via blockchain, then you are going to want to do everything in your power to ensure blockchain becomes as mainstream as possible.

It is important to keep in mind that it is likely going to be a tough road to hoe, however, as many of blockchain's greatest benefits are only going to be available to companies who already have the existing infrastructure to take advantage of them as fully as possible. As such, the most realistic forecast for the rise of blockchain technology is that there will be a handful of new companies that are going to come along and grab a share of the spotlight, while the rest of the room at the top is taken up by the members of the old guard who can get their acts together and make a move on blockchain technology before their competition has a chance to.

This is not a knock on blockchain technology; this is merely the way new technology is often assimilated into the mainstream. Remember, understanding this tendency is the best way to circumvent it and find the success you seek.

Chapter 5: How Blockchain Can Reshape Financial Services

Once they understand the benefits of blockchain technology, many people then make the mistake of jumping right to the question of how does it work. This is akin to asking how a microwave works when really you are likely curious as to what can be cooked inside of it. Considering what can be done with the technology is far more useful than understanding the specifics of how it works which will likely just get you further from the answer you are looking for.

As the key purpose of blockchain is in providing additional trust for a variety of transactions, which it does instantly and with a large volume of transactions, it is helpful to consider the areas of modern society where this will most frequently come into play.

Banking: A scenario where most banking transactions are replaced with blockchain transactions is the most commonly pointed to expression of blockchain technology in the modern world. While electronic payments have sped up the process of verifying banking transactions significantly, the amount of time that it can take for some transactions to appear in specific accounts can be a significant inconvenience, especially if the transaction takes place on the weekend.

The blockchain, then, removes the need for the wait entirely, making it possible for two people to complete a transaction directly, without ever involving a financial institution in the process directly. The transaction is checked by the blockchain, then verified by miners working on behalf of the blockchain, and then placed into the permanent record of the blockchain.

Digital currency: Bitcoin is the most commonly used example of the type of digital currency that blockchain technology makes possible. This, in turn, has led some people to confuse the two. If you have a hard time keeping them separate from one another, you can think of them the same way you think of email and the internet. Email (Bitcoin) could not exist without the internet (blockchain technology).

Asset exchange: Any asset that has a proven demand for claims related to authenticity of ownership could be improved with the addition of blockchain technology. As the cost of the blockchain is determined, and paid for, through its usage, the assets that are the most likely to receive blockchain are not necessarily going to be limited by value, but by the complexity of determining the chain of ownership, and thus the quality of the product. For example, a unique blockchain may not be required for the coffee at the corner store up the street from your apartment, as you can see where it comes from. On the other hand, if you were looking to buy your coffee in bulk you may seek out a blockchain to determine where the coffee came from and the quality of the beans.

Smart contracts: Despite the name, smart contracts aren't actually contracts, nor are they all that smart. They are simply small programs that are set to activate when a specific external event occurs. This means that while they are not legally enforceable, they can be added to a specific blockchain as per the agreement in a contract to work in tandem with a binding contract. For example, a smart contract could not legally require you to make your car payment on time; it could, however, connect to an Internet-enabled car and shut off the battery if the payment was not detected on time. However, it could only do this if you had already signed a contract saying the smart contract could be used.

Privacy: Another instrumental part of blockchain technology, in the current digital atmosphere, has to do with the collection of private data. Specifically, the fact that very little is ever available and what it can be tapered even more based on the desires of the person whose data is being displayed. At most, some relevant data is displayed to those that took part in the transaction, though even this is not guaranteed.

Commonly asked Blockchain questions

Will blockchain technology make all banks superfluous?

While blockchain technology is already proving to be disruptive, the fact of the matter is that it is far more likely to become a part of the traditional banking structure than it is to replace it completely. Many banks, including JP Morgan, are already investing heavily in ways that blockchain technology can be put to use enhancing, existing systems, not breaking them down. Furthermore, as national cryptocurrencies come online, it will become even less likely that a third-party service is going to come along and dramatically rewrite the status quo.

Will blockchain automation lead to a loss of jobs?

While every new technology tends to make certain types of jobs obsolete, the loss caused by blockchain automation is likely going to be minimal. Very little of the verification that a blockchain will be automating is still done by hand these days so that the automation will shift from one automated process to another. The financial sector is likely to be stricken in specific areas, but as blockchain is very much a financial technology, it is expected that they will be able to pivot to a new role in the blockchain economy. Additional jobs are also likely going to be added as new industries come into being to deal with the issues that a blockchain-based society may face.

Is everything stored in a blockchain 100 percent accurate?

While you can be sure that the version of the information that you see in a blockchain is accurate based on the information that the blockchain has been provided, this is very different than saying that this information is correct. Blockchains are going to be vulnerable not to hacking in the traditional sense, but what is known as social hacking instead. In this scenario, if enough of the nodes in a blockchain could be convinced that the information they contain is accurate when this is not actually the case, then those nodes could then bully the other nodes into picking up the inaccurate information.

If 51 percent of the nodes in a blockchain are hacked, then the entire blockchain can be compromised. What currently keeps this from happening is the associated costs, but eventually they will be less than the reward and when that happens social hacks of this nature will certainly occur.

Chapter 6: The Technology Behind Cryptocurrencies

A short history lesson

The first key aspect of what would go on to become blockchain technology was created in the early 1990s as a way of dealing with a new and dangerous threat, email spam. Known as the proof of work model, it is essentially a complicated equation that needs to be completed for a specific process to be executed. In the original usage case, this was used to ensure that a computer had to complete a proof of work to send an email. For one email this was no big deal, for a thousand or more, such as what a spammer would send out, it was a task that the computers of the day were not up to.

While the technology would not catch on for its original purpose, some ten years later it would be put to work in what today is commonly known as cryptocurrency mining. In this instance, it was a person or group of people using the alias of Satoshi Nakamoto who put the proof of work model to use in the opensource code for both blockchain technology and the Bitcoin cryptocurrency when they released the technology to the world in a treatise entitled bitcoin: A P2P Electronic Cash System in late 2008 on the networking website P2P forum.

The Nakamoto alias soon sent bitcoins to a handful of associates before mining the resulting transactions. At this point, many other programmers expressed interest in the technology and the Nakamoto alias faded into the background never to be heard from again. However, many people theorize that the alias left behind a bitcoin wallet, that, if found, would make the owner a billionaire.

In early 2009, one early adopter traded 10,000 bitcoins for two large pizzas worth about $20, setting the price of bitcoin at about $.002. As of December 2017, one bitcoin was worth as much as $20,000.

To understand just how blockchain functions, it may be helpful to understand just how bitcoin works as well. In general, it can be thought of as similar to other types of only payment systems, such as PayPal. The difference here is that the currency being exchanged has no true value outside of what the market says it is worth, there is nothing physical backing it up. This is one of the reasons the price for cryptocurrency is so volatile, there is no outside force working to keep the price somewhat stable.

Each of the transactions that are then processed through the system are processed to form the blocks of the blockchain that keeps Bitcoin up and running. Transactions are verified by individuals, known as miners, who are paid for their work by being awarded with 12.5 bitcoins per block that is mined successfully. This amount is then split between each of the miners who has a hand in verifying a particular block and helps to offset electricity costs in addition to ensuring a financial incentive for mining exists as well.

Miners use specialized computers, known as mining machines, to ensure that they can verify transactions as quickly as possible. Even still, the complexity of the average bitcoin proof of work that is required these days is far beyond what a single mining machine can successfully manage on its own. As such, bitcoin mining has become big business, with companies running massive server farms to keep up with the demand.

The number of transactions that can be verified per minute, as well as the rate at which new blocks are created, varies dramatically between different cryptocurrencies. Bitcoin is already pushing up against its upper limits which has led many analysts to theorize that a more well-equipped cryptocurrency could come along and steal its thunder.

In 2014, while Bitcoin was in the midst of its first serious boom, programmers discovered that they could add small amounts of complete code to blocks, which could then be activated when a specific criterion was met. Dubbed smart contracts, this addition to the blockchain pantheon has already gone on to inspire its own entire industry as well, along with Bitcoin's primary competitor, Ethereum. Smart contracts are useful in a wide variety of scenarios, from facilitating contract negotiation to tracking patients in a hospital and their inclusion in the blockchain bag of tricks has cemented it as a technology to watch moving forward.

Since that time, the cryptocurrency market has exploded, and the blockchain market has done the same. There are now more than 1,000 different cryptocurrencies on the market today, worth more than $60 billion when everything is said and done. Individually they are worth anything from several thousand dollars all the way down to less than a single cent. Along with this massive surge in interest has come an influx of new technologies that will be discussed in later chapters.

Chapter 7: Blockchain Applications That Are Changing Our Future

A new take on cloud storage: Studies show that approximately 70 percent of personal computer owners have 100 gigabytes or more of unused space sitting on their hard drives. Multiple companies, including Sotrj Labs, are striving to take advantage of this fact and make it accessible to those who need it through the power of blockchain technology. Their app takes advantage of the security inherent in blockchain technology to make sure that the files stored in this way are secure, while at the same time ensuring they are easily accessible as well.

Storj works as a sort of storage marketplace whereby it connects those who are looking for storage space, with those who have the extra space available. The peer-to-peer network that then connects the two is naturally faster than the large-scale alternatives, simply because it doesn't have to deal with heavy use that they do. As users can rent out precisely the amount of storage they need, the associated costs are also naturally going to be lower as well.

The way this type of service typically involves huge data centers that are prohibitively expensive to both run and build since server costs don't decrease when scale is factored into the equation. As such, building a thousand servers at once isn't going to cost all that much less than building them one at a time. These higher costs are then passed along to the consumer. Additionally, the fact that users' data is going to be stored in bits and pieces across thousands of different systems means that, when compared to a massive server farm, it is a much less tantalizing target for hackers. Currently, there are nearly 20,000 users on Storj, using storage on more than 7,500 different computers.

Singular online identity: While using a singular login identity across all of the internet seems like little more than a wonderful dream, a company named SecureKey has partnered with IBM to make it a reality. They are working on an initiative which is designed to promote this type of singular online identity on a global scale. When adopted in this way, it will have the ability to connect banks, telecommunications companies, healthcare providers and government bureaucracies through a single blockchain network that is accessible by the employees of each. This network is being built with the HyperLedger technology that was recently created by the Linux Foundation and will run on the back of an IBM blockchain.

When activated, it will make use of what is being called a distributed trust model which will allow users to determine which of their details are going to be shared, along with the organizations that are able to access them. It will also ensure that consumers will have to consent to the access of their information from each individual source.

Again, while it may seem as though this type of unifying system is still a way off, it is already seeing adoption by many of Canada's leading banks including the Canadian Imperial Bank of Commerce, TD Bank, Desjardins Group, Royal Bank of Canada, Scotiabank and the Bank of Montreal. The system is already robust enough to ensure that it will easily scale when new members are added to make it possible for them to get connected as simply, and effectively, as possible. Much like what is going on in the cloud storage space, the natural benefits of blockchain come into play in a big way when it comes to making this type of system an attractive package.

Smart contract applications: While you may not have heard much about them yet, smart contracts are already big business, just ask the Ethereum platform which is tied to ether, the second strongest cryptocurrency on the market today. This blockchain was created with the use of smart contracts in mind, and there are already a substantial number of decentralized applications that you can interact with right now.

Smart contracts are already seeing massive use via applications that do things such as take the place of a contract negotiator. They are also used in insurance applications. In this scenario, policies purchased in-app are then able to include smart contracts that connect internet-enabled vehicles to provide accident data to the central server which can then, unequivocally determine fault in an accident, without involving a person at any step in the process.

Smart contracts are also proving useful when it comes to making it easier for artists to keep track of uses of copyrighted works. One company, Blockphase, already has an application on the market that caters to artists working in the virtual reality, augmented reality and 360-degree space. Artists are then free to upload their content to the Blockphase blockchain where their technology then adds it to a database of content that is searched for on a regular basis in order to ensure that it is not being used without consent of the creator. If it is found to be in use, a letter is then automatically sent to the user requesting compensation.

On the commercial side of the digital media space, ContentKid is a company that is working to decrease the consumption of pirated media, by allowing users to pay for the amount of content that suits their specific viewing needs. Their service subscribes to all of the common content subscription services and then leases out access in 15-minute increments. This, in turn, gives users more options when it comes to their favorite content than pay a monthly fee or turn to piracy. Payment is all done fluently through smart contracts and the ContentKid blockchain.

Notary services: As many of blockchain's core features are naturally in line with what notary services offer, it is only natural that there are applications out there, including Stamped.io, which are officially providing these services. It allows users to access common notary services without going to the trouble of finding a bank who still staffs a notary on a regular basis. While this is more of a convenience in some countries, than others, in countries such as India, where the amount of red tape required for virtually any situation is extensive, this expression of the technology could be revolutionary. Along similar lines, the ease of use associated with this expression of the service will mark the first-time millions of people around the world have access to these types of services.

Apps like Stamped allow blockchain users to indicate the authenticity of a specific file from any internet browser. When a document receives this verification, it is then supplied with a digital timestamp with additional verification of the entire process coming in the form of a SHA256 hash on their blockchain. This process can be equally effective for virtually any file type and can be used to easily determine if a specific file was in the possession of a certain person at a specific time.

No more hanging chads: The issues that the US has experienced with their voting process in the twenty-first century means that there are plenty of people out there who are very interested in ensuring that blockchain technology starts revolutionizing the voting process as soon as possible. The reasons why should be obvious, starting with the fact that every vote that is cast, and verified, will be added to a blockchain which would make counting the votes accurately, down to a one, child's play. Each voter could also independently determine if their vote was recorded properly, making any institutional verification process unnecessary.

One app that is already up and running is known as **Follow My Vote**, and it is working to improve the transparency of elections of all shapes and sizes. Another benefit of this type of system is the fact that it would likely dramatically increase voter turnout because the process would be much less cumbersome, while also substantially reducing costs across the board.

Follow My Vote currently works with specialized voting booths that record personal details from voters, before adding them to the Follow My Vote blockchain. With this done, users can apply their voter ID number to vote from the app in any eligible election. A voter ID number would then give users the ability to watch the election play out in real time, and even change their votes at any point before the closing time of the polls.

Internet of things: IBM is also hard at work on a blockchain that will serve to connect a wide variety of disparate of home products that are capable of connecting to the internet of things. Through the Watson IoT platform, these devices will be able to communicate with one another for an, as of yet, undetermined number of purposes. Watson is also going to offer services such as text analytics, machine learning, natural language processing and more.

The idea of this type of connectivity is already having a serious effect on the automotive industry where the internet of things can be seen as an integral part of the smart car movement. In fact, more than 70 percent of all automotive manufacturers are already hard at work on integrating this technology into their future product lines.

Money transfers: While bitcoin is the leading name in the cryptocurrency payment space, several other contenders in the market are approaching the process in different ways. One of these is known as Ripple, which has created its own type of blockchain-based payment network that is connecting corporations, banks, payment providers and digital asset exchanges to make it possible to initiate a money transfer practically anywhere in the world. Ripple can already be found in more than 90 countries worldwide.

Ripple service works in three different ways. Banks make use of the xCurrent service to process payments on a global scale. At the same time, payment providers make use of a service known as xRapid to ensure that they have the liquidity required to complete trades regardless of the specifics. Finally, businesses can use the xVia service to send payment directly via the Ripple blockchain. Ripple also uses its cryptocurrency, XRP, that is already listed on numerous different cryptocurrency exchanges.

Chapter 8: Technical Guide to the Blockchain Technology

Decentralized database: When compared to a more traditional database, the biggest difference with a decentralized database is the way it deals with lag. With a centralized database, ensuring the data can be accessed as quickly as possible is the primary goal, and all of the servers are stored as close to one another as possible to facilitate this fact. When it comes to a decentralized database, however, all thoughts of defeating lag are thrown to the wolves in favor of creating a database that can be accessed from literally anywhere in the entire world. When this facet of the technology is combined with its already impressive security and auto-sorting functionality, it creates a database experience like no another.

Hashes: Data that is stored within a blockchain can be broken down into two types, transaction data and data that provides the block with a specific location on the blockchain and uniquely identifies it. While the first type of data takes up a vast majority of the block, the data about the block itself is nevertheless extremely complicated. Once the transaction data is verified in a block, it is then encrypted through what is known as a hash function. With the data encrypted, even if it is stolen by a hacker it will be completely useless without the relevant decryption key. What's more, the hash for each block can be thought of as a type of digital fingerprint, as such, if the data is compromised, the hash will be changed and the blockchain will take note and remove it before it can spread.

The most commonly used hash function is the SHA256 hash, data encrypted with this type of hash function can only be decoded with the compatible hash key. Each block receives its own hash when it is added to the chain, but that is just the start of the hashing fun. Each new hash is then also taken into account for the hash for the blockchain as a whole which means the overall hash is updated each time a new block is added to the chain.

Merkle Tree: The reason for the existence of the blockchain level hash has to do with what is known as the Merkle process or a Merkle tree. This process ensures that the blockchain is able to work as effectively as possible while still maintaining an overall degree of accuracy. The Merkle process is essentially a functionality matrix that serves to maximize efficiency. They also make it easier for financial transaction to be tracked to make it possible to follow otherwise complex strings of transactions.

While a Merkle tree isn't a required part of a blockchain, those that don't include them tend to process transactions less efficiently, while at the same time proving to be less secure as well. This is because the blockchain level hash, also known as the root hash, can be viewed as the sum total of everything that is currently a part of the blockchain. This means that when the Merkle tree goes to work verifying the current sanctity of the blockchain it doesn't need to worry about looking at every transaction specifically, and instead only needs to determine that the root hash is intact because any inaccuracies in the blockchain as a whole would be reflected in the root hash as well.

The Merkle tree received its name based on the way it works, specifically, if it comes up against a potential issue it creates a branch at that point, and then continues to check the blockchain assuming that the issue is valid and invalid at the same time. This, in turn, ensures that the sum total of the blockchain is always verified as quickly as possible in order to keep things moving briskly at all times.

Checking the amount of data in the average public blockchain manually would be virtually impossible due to the time-consuming nature of the process. In addition to making this process manageable and automatic, the Merkle tree also limits the amount of data that needs to be shared between nodes to just the root hash as opposed to all of the transaction data. Then, as long as 51 percent of the nodes agree that things appear to be on the level, the Merkle tree considers its results valid and then restarts the process with the next new block.

To ensure that a Merkle tree is able to run as smoothly as possible, it is important that a high level of trust is maintained at all times when it comes to the infallibility of the system as a whole. If this level of trust is not maintained at all times, then the likelihood of the blockchain's viability continuing in the long-term becomes far more uncertain. This, in turn, allows any users who are interested in setting up their own nodes to download the latest version of the software with the confidence that they are downloading the most accurate version of the blockchain that is currently available.

Furthermore, there needs to be trust between the various node operators that the Merkle tree will able to pinpoint and eliminate any false nodes that are set up by pointing out scenarios where certain hashes don't match up the way they are supposed to. As long as the Merkle tree is functioning properly, then multiple nodes can send back the same false information and it will all be flagged appropriately before being replaced with the last verified accurate version of the blockchain that is available.

Chapter 9: Smart Contracts

When it comes to making the most out of smart contracts, the front lines these days are on the Ethereum Platform. While the Bitcoin platform is, first and foremost, a means of facilitating cryptocurrency transactions, Ethereum is heavily focused in the area of smart contract technology and a majority of its applications run some type of smart contract or another.

The Ethereum platform was created by a man named Vitalik Buterin while he was working as a programmer for Bitcoin. In 2013, Buterin had the idea of creating a scripting language for the Bitcoin blockchain that would allow it to run decentralized applications. His superiors, perhaps correctly, were afraid that the blockchain couldn't handle the additional functionality and his proposal was denied. Rather than give up on his dream, Buterin decided to leave his job and started working on the Ethereum platform full time in the spring of 2014 before opening it up for use the following summer.

Mapping: A mapping is a type of associative array that associates balances and addresses. These addresses are then stored in the common hexadecimal format, while the balances are converted to integers ranging somewhere between 0 and 115 quattuorvigintillion. If the mapping includes the public keyword, then the variable in question will be visible by everyone on the blockchain. This is crucial in order for clients to display them properly.

MyToken: If you published your contract immediately, it would attempt to use available coins, but at the same time would be unable to find any. As such, in order to make the contract useful, you need to create a few tokens as well. This can be done with the MyToken function by changing the balanceOf variable to something greater than zero. The function is based on the contract MyToken which means that if you change the name of the contract you will need to change the name of the function in order to ensure things continue working properly. This function then sets the balance to the determined amount in the account of the user who deployed the contract in the first place.

When you are creating your smart contract in the Ethereum platform, you can accomplish much the same thing from the drop-down menu on the right hand side of the screen. There you will find the "pick a contract" option which holds a option for MyToken. Within that option you will then find a section known as constructor parameters which offer numerous parameters for your token including the initial supply and the transaction fee.

Transfer function: The transfer function is very straightforward, it has a value that serves as a parameter as well as a recipient and when it is used it then subtracts the relevant _value from one balance and adding it to another. The biggest issue that needs to be addressed from a coding perspective is the way in which users are prevented from sending more _value than their existing balance allows for. To mitigate this issue you will need to include a check that guarantees the sender has enough funds to execute the transaction otherwise it is stopped. This method also serves to prevent overflows which will prevent a number from getting so large that it rolls back over to zero.

A contract can be canceled mid-execution through the use of either a throw or return. A return tends to cost less but is also more of a hassle as it cancels out of the contract but keeps any changes that the contract received in the interim. As such, the throw command is typically cleaners as it reverts changes to transaction that could have been made, though it does cause the user to lose any gas they have put into the transaction. It will also generate a warning that this will occur to the user, however, making the loss on them if they go through with it.

Deploying a contract: Once you have completed your contract, you will need to deploy it in order for it to do much of anything. To do so, you will want to use the token source code outlined above, open your Ethereum Wallet and then paste the code into the source field for Solidity. Assuming the code compiles properly, you will then see the option to pick a contract from the right drop down menu. You will then be able to select the contract labeled MyToken. You will then be given the option to personalize your token a little more.

At the bottom of the screen you will be presented with an estimate as to the cost of the contract you are creating, in general the default settings should be sufficient. Pressing deploy will set tings in motion and ensure you are redirected to the front page where you should see your transaction waiting to be confirmed. Choosing the name of your primary account will then transfer 100 percent of the shares of the new token to that account.

You will then be able to send your currency to anyone whose Ethereum wallet address that you have access to as any Ethereum wallet will naturally be able to store tokens created through the blockchain as well. This does not, however, mean that the wallet will detect the new cryptocurrency automatically, however as this detail has to be added in manually. It can be found under the tab labeled Contracts where a link should be visible that will take the user to the freshly minted contract. You will then want to copy the address into your text editor.

To add the token to the wallet's watch list, you will go back to the contract page before selecting the option to Watch Token. This will present you with a pop-up box that will appear, into which you will need to paste the contract address. This, in turn, will fill in the name of the token, its symbol and its decimal number. Any details changed here will only reflect how the token is viewed in the wallet, not anything about the currency itself. With this done, the wallet will then display the number of tokens that are currently available and allow interaction with them as normal.

Chapter 10: Blockchain Technologies and Cryptocurrencies to Watch

Zcash: Zcash is a type of cryptocurrency that looks like any other on the surface, if you dig a little deeper, however, you will notice one significant difference. Specifically, its transactions are even more private than what other types of blockchains offer. You see, while those outside of the transaction can't see any of the identifying details of any of your transactions, a record of those transactions still exists on the blockchain. This, in turn, means that with sufficiently advanced software it would be possible to have a fairly good idea of who initiated which transactions.

This is not the case with Zcash, however, as it makes use of what is known as zero-knowledge cryptography which allows transactions to proceed while at the same time masking their precise specifics. These anonymous transactions are then referred to as being shielded by Zcash. Without this type of protection, it is possible for not only anyone to determine the transactions details of a specifically known wallet, but also the wallets of everyone that wallet has ever transacted with. Even if you have nothing to be ashamed of by your transactions per say, it doesn't take much to see why an improved on this system wouldn't hurt.

Like Bitcoin, there is also a total of 21 million Zcash in circulation. Unlike Bitcoin, Zcash's further development is funded directly by use of its blockchain. Specifically, 20 percent of all of the Zcash mining rewards are going to go to the stakeholders of the Zcash company for the first four years of the company's existence which will total 10 percent of the overall supply.

Specifically, the technology at play here is known as zk-SNARKs or zero knowledge succinct non-interactive arguments of knowledge. This technology allows Zcash users to determine how much of their private information is disclosed along with each transaction. This allows users to still have the option of providing their information when it is useful for them to do so.

Moreno: Moreno is another type of cryptocurrency that focuses on privacy above everything else. Started in 2014, it works towards the same goals as Zcash. The cryptocurrency of Moreno XMR, is worth about $350 as of December 2017. Unlike Zcash, which simply doesn't keep track of the details it is told not to, Moreno adds additional layers of security to the mix, starting with ring signatures.

Ring signatures hide the details of where transfers originate placing each true transaction in with a number of other transactions that appear real to an outside eye. This is coupled with the use of stealth addresses which hide where the money is being sent by being randomly generated at the time of the transaction.

Finally, ring confidential transactions obfuscate the amount of the transaction in question. This is done by forcing each person who sends XMR to enter an amount for the transaction as well as the amount they want returned in change. This then allows the transaction to be verified by ensuring that the sum of the outputs remain the same and that no new coins were created as a result. The true amounts are never shown, however, with the amount itself being cryptographically hidden while also including a range of proofs to ensure things proceed as anticipated.

While some believe that cryptocurrencies of these types are only going to help cryptocurrency return to its darknet roots, the fact of the matter is that there are plenty of perfectly legitimate enterprises that have good reasons for not sharing their transaction history with the world. Merchants of all types have a vested interested in keeping their details hidden as they clearly do not want everyone to know exactly how much they are buying and selling their stock for. In fact, as blockchain technology becomes more and more prevalent, it is likely that the spotlight will fall on this technology once merchants realize it is a problem that needs solving.

In order to ensure their technology actually gains traction among mainstream cryptocurrency users, the website XMR.to has been created as a means to bridge the gap between Moreno and Bitcoin. The website allows anyone with a Moreno wallet to send money to any Bitcoin address, despite the fact that they operate on different blockchains. Not only does this make it easy for XMR holders to use their cryptocurrency to complete transactions with millions of individuals around the world, it allows them to do so in such a way that the transaction doesn't show up directly linked to either party on the bitcoin blockchain. As rumor has it the US government has started looking into those who are not paying taxes on their bitcoins, this process might become even more popular as tax time comes around.

Dash: Unlike many other altcoins, Dash doesn't want to do anything other than act as an alternative to fiat currency. Built on the bitcoin blockchain, with improvements to privacy and transaction speed thrown in, it was worth about $1,000 per unit at the end of December 2017 with a total of 18 million coins. Dash has a variable block reward which decreases at a 7.1 % rate each year. The average block mining time is 2.5 minutes on the Dash blockchain, which makes it four times faster than Bitcoin.

What sets it apart from other types of cryptocurrency, however, is that its transaction fee is negotiable.

Features of Dash include the ability to send a transaction as private (worth up to 1000 dash) by mixing your transaction in with numerous others being transacted at the same time. More interesting, however, is its instant send feature which sends transactions instantly, though high transactions fees apply. This is done by sending prioritized transactions to a set of masternodes that set outside of the traditional blockchain and stand ready to complete transactions virtually as fast as they are started. In order to be listed as a masternode, users need to own a minimum of 1,000 dash coins, when new coins are created 45 percent go to masternodes and 45 percent go to miners with the rest going to fund development.

Additionally, 10 percent of all of the mining rewards from the dash mining process are set aside to be returned to the creators for the purpose of improving the cryptocurrency even more. Currently, much of this fund is being used to develop a use of a new point of sale software that is being targeted at emergent industries in the US that are looking for alternatives to traditional financial structures. Dash also allows its users to vote in order to determine what projects are going to be focused on next.

Hyperledger: This is not so much a new project as an ongoing opensource program that is being developed as a way for developers to interlink the blockchains of several different companies quickly and easily. When it has been completed it will be a type of global collaboration platform that will be used to connect virtually anything that is already activated in the internet of things. The project is being sponsored by the Linux Foundation as well as members of the supply chain, technology, manufacturing, financing and banking industries.

What Hyperledger represents is the next level of blockchain implementation which will likely act as the basis for a wide variety of the blockchain applications of the future. The success of this project is mainly based around the contributions of the developer community funded by the backing of the major industries who have a vested interest in seeing the technology take off in a big way. Due to its open sourced nature, anyone who is interested in making this particular dream a reality can get in touch with the Linux Foundation to determine the best way they can help out.

Currently, there are more than 100 members working with the Linux Foundation to improve what Hyperledger can do. The list includes industry leaders such as Samsung, Intel, Nokia,Huawei, SAP, Fujitsu, IBM, Daimler, Airbus, Deutsche Börse, J.P. Morgan, American Express and more.

Current projects include

- **Hyperledger Cello:** Hyperledger Cello aims to bring the on-demand "as-a-service" deployment model to the blockchain ecosystem to reduce the effort required for creating, managing and terminating blockchains.

- **Hyperledger Sawtooth:** Hyperledger Sawtooth is a modular platform for building, deploying, and running distributed ledgers. Hyperledger Sawtooth includes a novel consensus algorithm, Proof of Elapsed Time (PoET), which targets large distributed validator populations with minimal resource consumption.

- **Hyperledger Iroha:** Hyperledger Iroha is a business blockchain framework designed to be simple and easy to incorporate into infrastructural projects requiring distributed ledger technology.

- **Hyperledger Fabric:** Intended as a foundation for developing applications or solutions with a modular architecture, Hyperledger Fabric allows components, such as consensus and membership services, to be plug-and-play.

Chapter 11: Business in the Age of Blockchain

Businesses, especially small businesses, have long had a love/hate relationship with the traditional banking establishment. On the one hand, it has always been required in order to ensure a variety of transactions are verified and secured and, on the other, they have routinely proven time and again that they are not out to do any business owners any favors.

As such, businesses that seek out more details on various blockchains will likely find that they not only offer more reasonable fees, but a wide variety of additional benefits as well. As an added bonus, the worldwide nature of most cryptocurrencies means that by adopting this type of approach many businesses could significantly expand their potential client base.

Another important facet of the technology to consider in this context is the distributed ledger aspect of the blockchain and what having access to transaction data at that granular of level is going to mean for most businesses. This information isn't going to be provided at the whim of some third party either, it is naturally, and readily, available to anyone who is a part of the transaction in one way or another. Even better, it allows business owners to remain completely in charge of their own personal data which makes a variety of cyber security attacks far less likely than those against large, collective, datastores.

Whenever a transaction is processed via the blockchain, the fee that is charged is split between the blockchain itself, to pay for infrastructure maintenance, and the mining reward, with the rest going to cover part of the reward that will be paid out when the block as a whole is verified. Nevertheless, a vast majority of the time the fees that the blockchain charges are going to be far less than what a bank would charge for similar services. When the average number of transactions for most businesses are taken into account, the resulting savings could prove to be quite substantial.

Additionally, dealing with transactions that are already on the blockchain is far easier when it comes to things like settling or clearing bank transactions. While these processes can take days or more with traditional establishments, the same level of verification can happen on a blockchain in minutes, if not even less. Nine times out of 10 a blockchain transaction is going to verify faster, regardless of the specifics, especially if the transaction takes place over the weekend.

When these ease of use benefits are attached to smart contract technology, the possibilities associated with both automation and streamlining all sorts of business practices become much more elaborate. While some of these possibilities are going to be out of the reach of smaller businesses, the cost of implementing blockchain technology is only going to decrease as doing so becomes more commonplace, and the tasks required to maintain them become more mundane. All in all, given enough time this type of technology is virtually guaranteed to make any type of organization run more efficiently while at the same time reducing operating costs across multiple departments.

Much of this will come about as a result of the streamlining that will occur when it comes to dealing with contracts on a regular basis. As a vast majority of all business involves transferring value between two or more parties, it stands to reason that implementing a usage scenario that takes advantage of the most secure environment for doing so (on a blockchain) that is currently available to the public. As an added bonus, any contract that takes advantage of blockchain technology can ensure all of its if/then statements are properly followed through on by attaching them to smart contracts that are guaranteed to follow through if, and only if, a specific activating event occurs. Even better, these actions take place in real time.

This powerful mixture of code and contracts is unique in that it allows every party in every business transaction to operate with complete faith in the system, and as a result, the other parties in the transaction. This level of faith will improve relationships with vendors and subcontractors at all levels of business. Smart contracts are also sure to prove useful when it comes to fighting against corporate bloat due to the layers of red tape, and lawyers, that will no longer be required in order to ensure that things play out the way they are expected to.

It is also sure to cut down on the costs associated with going after those who don't follow through on what their contracts specify. This is due to the fact that once an activating event has occurred, there is no way for a smart contract not to follow through on its programming. Once the specifics are set, and the block has been added to the blockchain, there is no going back.

Chapter 12: Executive Guide to Implementing Blockchain Technology

If you are interested in ensuring that blockchain technology is implemented in your company successfully, it is crucial that you approach the problem with the goal of making the most of the strengths your company already has, rather than looking to use blockchain technology as an excuse to pivot in a new direction. As previously noted, the technology as a whole is still in its infant stages which means it is definitely not in a state where you want to completely restructure an existing company around its use.

Rather, you are going to want to look to ways to invest that will allow the company to approach the new technology in a strategic way that will serve to minimize costs while maximizing benefits at the same time. The best way to approach this problem, then, is to create a core technology working group to work out the most effective way of implementing blockchain technology into the core competencies of the company. It is important to leave someone in charge who has the best interests of the company in mind, however, as it can be easy for these groups to get so caught up in their new technology that they try and include it in as many different places as possible, regardless of what the practical benefit might be.

The person overseeing this working group is often known as the blockchain czar and is generally from middle management. While the rest of the group is working to theorize the ways that blockchain can be added to everything under the sun, the czar will be the one in charge of making sure that the blockchain project retains a viable cost benefit analysis that focuses on relevant strategic goals. In general, the czar will be in charge of orchestrating the flow of the conversation among the working group, and also during the meetings that the working group has with other aspects of the company.

Getting started: In order to ensure that the conversation proceeds in a productive fashion right from the start, the first thing the working group will want to do is consider the blockchain projects that will present the technology in the best light to the company, making it easier to greenlight other useful, but less flashy blockchain projects in the future. The best place to go when it comes to adding ideas to this list is going to be the pain points that the company traditionally deals with. Nothing immediately shows the value of a new technology like replacing a commonly used workaround or clearing a common client complaint all at once.

While having an initial list in hand is beneficial, it is important that the final decision as to how to more forward isn't made without receiving input from relevant shareholders and specialists both internal and external. Remember, the working group is only going to have one chance to make a pitch for its continued existence, and a lackluster initial showing is rarely going to create enough interest to get the first, much less the second or third, blockchain project off the ground.

It is common for work groups during this stage to get caught up in choosing projects that cause the greatest amount of disruption overall or those that adhere most closely to things that are popular in the moment. It is then going to be up to the blockchain czar to operate as the voice of reason, keep the big picture in mind and remind the group that creating institutional change is not the work of a single project.

Hopefully, this will, in turn, make it easier for the team to zero in on projects that are going to do the most to improve the capabilities of the company in question in the shortest amount of time possible. Ideally this will include ways in which they can help the company outpace the competition or to otherwise generate proof that blockchain technology is worth backing more aggressively in the future.

Form a hypothesis: After the best starting point has been determined, the next thing the working group will need to do is develop a hypothesis in regard to how the blockchain technology, once implemented, will actually make a difference to the company. For example, if the hypothesis is that a distributed ledger would be an effective way to keep track of otherwise disparate transaction information, then they would need to come up with a means to test whether this is a true statement in the long run.

Putting the hypothesis into play will also require additional consultation from relevant parties who, at this time, may include internal and external business groups, function teams, relevant customer segments and other stakeholders. The working group should also contact important inter-organizational groups including regulatory committees, compliance teams, operations, IT and financing, among others. While this amount of buy-in is certainly going to make getting the initial project off the ground a much more complicated process than it would otherwise be, it will also ensure that the project that is ultimately pursued will have the ability to go the distance when it matters most.

Moving forward: After the working group has put together an actionable hypothesis, the next thing that will need to be done is actually putting that hypothesis into play. When it comes to the implementation of the prototype, it is perfectly natural for them to adjust the parameter until the prototype is working as effectively as it possibly can. As such, during the testing and evaluation phase the team is inevitably going to come to a consensus when it comes to things like common practices and also uncover new, and potentially unexpected, ways to put the blockchain technology to work as effectively as possible.

During this phase of the process, it is also perfectly normal for a prototype to become so altered that it no longer reflects the original hypothesis. If this is the case it is then the blockchain czar's job to determine if a realignment with the original prototype is in order or if something about the new prototype warrants a reconsideration of the hypothesis instead. As long as the end result remains a good example of putting the company's core strengths to work and meets the original goal the team set out to master, then the particulars may not be as important.

It is also the czar's job to pay attention to the momentum of the project overall as just because the technology is new, and the implementation is complex, is no reason to consistently fail to hit milestones. If the old milestones no longer make sense, it is czar's job to figure out why and how they need to be altered, if appropriate. Likewise, when testing, it is going to be the czar's job to ensure that the testing constitutes fair practices and that they aren't biased towards a pro-blockchain result. This, in turn, means that the time given for development should be long enough that the team can create something worthwhile, while not spooling out for so long that the goal ends up getting lost in the shuffle.

Rolling out: After the prototypes have proved their success, the last thing that will need to be done is to run the numbers when it comes to implementing the new technology on a company-wide scale and determining the cost of doing so once and for all. Prior to implementing the recommendation from the working group, it is important to take into consideration the ways in which the recommendation will support the core functions of the business and, ideally, impact them for the better. The final proposal should make it clear the way in which blockchain technology will change the way the company does business either with its customers or its vendors.

Furthermore, it is also important to keep in mind that a full-scale implementation of even a fairly straightforward blockchain implementation is going to be a long-term project. As such, the early goals should be related to things like improving compliance, reducing costs and improving overall quality and the like, and then get buy-in on them from all the relevant parties. A unified approach will make it easier to create an implementation and scaling roadmap that actually has a chance of being followed.

The roadmap that is being followed should also then include the ways in which the company will be affected by the new technology, the costs associated with it and what the estimated benefit is going to be. If this roadmap doesn't make sense from a financial perspective, then it is important to not throw good money after bad and write off the entire venture as a learning experience. On the other hand, if things end up working out from a financial sense, then you know exactly what to expect moving forward.

Decentralized application creation

If part of the plan for your company involves the creation of an application that will interact with the blockchain specifically, then you are going to want to use the Ethereum platform to make this dream a reality. Ethereum is the name in blockchain-based app development these days and many developers have even switched from other blockchains because the benefits that Ethereum brings to the table are fairly substantial. In order to create an application, your company will need someone who understands the Solidity programming language, which uses, .sol and .se extensions along with LLL a lisk byproduct. If a Solidity programmer is not available, someone who knows Python and Serpent will likely be able to find their way around.

When it comes to compiling programs, a C++ solc complier is required; along with the Web3.ja API if your application is going to need to connect to a smart contract directly with the use of JavaScript code. This is a very important step if you want your application to be accessed by those who aren't running their own Ethereum nodes directly.

In order to get things up and running as quickly as possible, you will likely want to go ahead and use a distributed application framework that is prebuilt, instead of paying to have a new one built from scratch. The Ethereum community is positively lousy with developers who are interested in promoting the use of the service as a whole which means there are plenty of quality APIs to choose from.

When it comes to the widely agreed upon best stack possible, the Meteor framework is the place to be, which is useful as it natively works with the Web3.js API as well which makes it the first choice of many developers working in the space these days. The meteor framework is also extremely supportive of Ethereum overall, which means new improvements are always being rolled out as a result.

You will also want to look into both Embark and Truffle which will help to streamline the overall process of app production. Truffle is great for companies that are getting into the blockchain business for the first time as it automates many parts of the programming process which, in turn, means that programmers will have an easier time of finetuning the code they create as opposed to going over the same basic steps time and again. Embark, on the other hand is sure to come in handy when it comes to automating the testing process which will help to keep the flow of the project moving as smoothly as possible.

The API that you will encounter most frequently is available from BlockApps.net and will act as Ethereum node for the purposes of interacting with any apps that are created. Another, similar, service is called MetaMask and it makes it possible to run an Ethereum node from virtually any web browser. If your company's implementation is going to involve several different user types, then you are also going to want to look into LightWallet which makes it possible for decentralized applications to provide different interfaces to different classes of users.

Creating an application

In order to ensure a decentralized application gets up and running successfully, the first thing that needs to be done is the generation of an Ethereum node that it can be accessed from. This can be done directly from the Ethereum website by downloading Geth, the Ethereum interface. It can be downloaded from Install-Geth.Ethereum.org. Once you have installed Geth you will then be able to open the programming console directly. Once it is open, you can quit out by pressing the ENTER key. When this happens, the console will then log the last thing you were working on so that it can then be retrieved at a later date using the command tail-fgeth.log.

Once this is completed, you will then be able to create both smart contracts and applications. The limitations on what is possible in this context is constantly shrinking so the odds are good that if you can envision it then you can make it a reality. After you have finished writing the code, you will then want to compile your work with the help of a solc C++ compiler. Once this is completed, you will then need to pay a gas fee in ether and provide your digital signature before proceeding.

With that out of the way, you will then receive the URL that corresponds to your application or smart contract's place in the blockchain along with the relevant API details as well. Once this is received you will then be able to access the application from virtually anywhere, not just when directly connected to an Ethereum node. Some applications will require gas to power them for each usage while others will not, it will all depend on the amount of data the application needs to pull from the blockchain on a regular basis.

Once any testing has been completed, the contract will then be ready to deploy via Truffle. To do so, all you need to use is the command truffle init to create a new primary directory for the application. You will also want to make sure you test compile before moving forward to ensure things are going to proceed smoothly. With this out of the way, you will then need to locate the contact that you have created and add it to the save file for contracts found in the folder located at app.json/config. You will then need to launch your node using the tesrpc command. Finally, you will want to relaunch Truffle and select the option to deploy from the root directory.

Additional functionality to consider

Mapping: A mapping is a type of associative array that associates balances and addresses. These addresses are then stored in the common hexadecimal format, while the balances are converted to integers ranging somewhere between 0 and 115 quattuorvigintillion. If the mapping includes the public keyword, then the variable in question will be visible by everyone on the blockchain. This is crucial in order for clients to display them properly.

MyToken: If you published your contract immediately, it would attempt to use available coins, but at the same time would be unable to find any. As such, in order to make the contract useful, you need to create a few tokens as well. This can be done with the MyToken function by changing the balanceOf variable to something greater than zero. The function is based on the contract MyToken which means that if you change the name of the contract you will need to change the name of the function in order to ensure things continue working properly. This function then sets the balance to the determined amount in the account of the user who deployed the contract in the first place.

When you are creating your smart contract in the Ethereum platform, you can accomplish much the same thing from the drop-down menu on the right-hand side of the screen. There you will find the "pick a contract" option which holds an option for MyToken. Within that option, you will then find a section known as constructor parameters which offer numerous parameters for your token including the initial supply and the transaction fee.

Transfer function: The transfer function is very straightforward, it has a value that serves as a parameter as well as a recipient and when it is used it then subtracts the relevant _value from one balance and adds it to another. The biggest issue that needs to be addressed from a coding perspective is the way in which users are prevented from sending more _value than their existing balance allows for. To mitigate this issue, you will need to include a check that guarantees the sender has enough funds to execute the transaction otherwise it is stopped. This method also serves to prevent overflows which will prevent a number from getting so large that it rolls back over to zero.

A contract can be canceled mid-execution through the use of either a throw or return. A return tends to cost less but is also more of a hassle as it cancels out of the contract but keeps any changes that the contract received in the interim. As such, the throw command is typically cleaner as it reverts changes to the transaction that could have been made, though it does cause the user to lose any gas they have put into the transaction. It will also generate a warning that this will occur to the user, however, making the loss on them if they go through with it.

Deploying a contract: Once you have completed your contract, you will need to deploy it in order for it to do much of anything. To do so, you will want to use the token source code outlined above, open your Ethereum Wallet and then paste the code into the source field for Solidity. Assuming the code compiles properly, you will then see the option to pick a contract from the right drop down menu. You will then be able to select the contract labeled MyToken. You will then be given the option to personalize your token a little more.

At the bottom of the screen you will be presented with an estimate as to the cost of the contract you are creating, in general the default settings should be sufficient. Pressing deploy will set tings in motion and ensure you are redirected to the front page where you should see your transaction waiting to be confirmed. Choosing the name of your primary account will then transfer 100 percent of the shares of the new token to that account.

You will then be able to send your currency to anyone whose Ethereum wallet address that you have access to as any Ethereum wallet will naturally be able to store tokens created through the blockchain as well. This does not, however, mean that the wallet will detect the new cryptocurrency automatically, however as this detail has to be added in manually. It can be found under the tab labeled Contracts where a link should be visible that will take the user to the freshly minted contract. You will then want to copy the address into your text editor.

To add the token to the wallet's watch list, you will go back to the contract page before selecting the option to Watch Token. This will present you with a pop-up box that will appear, into which you will need to paste the contract address. This, in turn, will fill in the name of the token, its symbol and its decimal number. Any details changed here will only reflect how the token is viewed in the wallet, not anything about the currency itself. With this done, the wallet will then display the number of tokens that are currently available and allow interaction with them as normal.

Chapter 13: Industries that Blockchain could Transform in the Future

Transportation: As ride sharing services become more and more prevalent, it is only natural that blockchain technology be put to use to further streamline an already efficient service. Specialized ride sharing blockchains are likely going to become a part of the private transport industry, which will, in turn, lead to the creation of specialized ride-sharing cryptocurrencies and all the ancillary creations that will entail.

Funds will then be deposited into specific, ride-sharing accounts, likely from numerous different types of cryptocurrency, and then these funds will be automatically transferred to the account that the driver has linked their smart car to. This will ultimately ensure trust as drivers and riders will both always know that fares are being accounted for correctly. Smart contracts can then be set up to ensure that new funds are transferred into the rideshare blockchain whenever the balance hits a specific level.

The added ease of use, to a service that is already extremely simple, means that these services are likely going to continue proliferating at an ever-increasing rate. Instead of being owned by a private company such as Uber or Lyft, assuming the rideshare blockchain is public, anyone could connect to it and offer anyone else a ride. Both parties could still feel secure as the start of their transaction was timestamped and the location of the node that processed the transaction can be traced as well should the need arise. The GPS information that is provided by the IoT connected vehicle would also be included, making the entire transaction completely transparent.

Another variation on this system is already in development that will allow IoT vehicle owners to do things like pay for fuel, tolls and even any tickets that are earned while the car is on the road, all automatically deducted from the prespecified account. While this sort of system would need to be adopted on a mass scale in order for it to be effective, the fact that it is being seriously considered shows the direction that these types of services are likely to proceed in.

Security: Remaining secure in the age of the internet is an ongoing problem without any readily available solutions; so much so that the World Economic Forum lists it as one of the most persistent threats facing the world today. Blockchain's previously stated level of security makes it a natural choice to help deal with this issue which is why it is likely to start showing up in these types of conversations with increasing frequency in the coming years.

While the security features that come part and parcel with blockchain can certainly be used to improve numerous different security systems as is, perhaps the most significant cybersecurity boosts could come from the use of blockchain to rebuild the internet as it is known today from the ground up. While this certainly sounds like a tall order, blockchain's decentralized nature could easily be adapted to mitigate the potential effectiveness of things like distributed denial of service attacks. Likewise, if the new internet properly took advantage of this type of decentralization it could make many similar types of issues cease to exist in the first place.

If built properly, it could also be used to dramatically improve the way that personal information is handled across the internet. Assuming the right standards were set in place at the start, users would easily be able to control just how their information appeared online and also where, and more importantly where not to make it visible. Each person would have a unique cryptographic key assigned to them where they could then store all of their information to be doled out automatically as the need arises. Information could easily be stored on a website in a single-serving fashion so that it was automatically removed once the site was done with it.

The music industry: The music industry is likely poised to experience the biggest shakeup since the creation of the mp3. Luckily, this next shakeup is likely to prove as beneficial as the last one proved destructive. To understand how this space is likely to evolve, it is helpful to look to the startup companies that are already hard at work looking for ways in which a blockchain can be inserted into the process of buying and listening to music. As an added bonus, it is also poised to improve the ways in which both record labels and artists receive royalties for the art they generate.

This usage scenario will utilize blockchain to make it easier for rights holders to seek out instances where their work is being used in an unauthorized fashion and also make it easier for those rights to be acquired automatically with the help of a properly configured blockchain that takes advantage of smart contracts to charge for content when illegal uses are found. When fully up and running the system should allow for artists to remove the middle man of distribution completely and instead be able to reach out to their fans directly and make money based on the relationships they cultivate.

Medical field: The healthcare sector is currently experiencing a serious shakeup as well thanks to the benefits that blockchain technology can provide when it comes to keep track of individual patients. Early tests of blockchain-based records keeping shows that using a blockchain to ensure each patient stays with their proper data can reduce the likelihood of mistakes by more than 30 percent on average and by greater than 50 percent during an emergency.

Aside from transmitting file, blockchain technology is also proving it can make a real difference when it comes to the ease with which hospitals and other medical practitioners store and share data. In addition to providing a much smoother experience for patients who are traveling between different caregivers to deal with the same issue, it will also make it much more difficult for hackers who target this segment of the market. It will also allow for more effective and all around faster diagnosis of numerous illnesses that the lack of interagency communication inherent in the current system makes difficult if not impossible.

There are numerous different startups in this space already, including Gem in California and Tierion in Connecticut. These companies are working on more than making medical records readily available, they are working to create a blockchain that everyone, everywhere is always connected to, making pertinent personal details only a click away while at the same time ensuring that those that are private remain that way no matter what. It could even be used to track medical test subjects, monitor their vitals and then, with the help of smart contract technology, pay them for their time when their services are no longer required.

Data storage: While cloud storage is an ever-increasing market, there are still several crucial challenges that need to be worked out in order to make it a worry-free alternative for corporations. Blockchain technology has the tools to deal with these sorts of issues, starting, first and foremost, with security. While the inherent security benefits of the blockchain would be a plus, the biggest boon to security in this scenario is that businesses would be able to maximize their existing available storage without having to worry about sensitive material leaving company-controlled digital space. The amount of underutilized storage tends to scale with the number of employees a company has, meaning that this has the potential to unlock countless terabytes of extra storage for major corporations, the equivalent of millions of dollars saved in the long run.

Supply chain logistics: The current ideas surrounding supply chain management are likely going to experience a serious series of disruptions in the coming years, thanks in large part to blockchain technology. This is based on the simple fact that both tend to focus on recording transactions in the best way possible, and blockchain is the far more efficient and easier to use of the two. So much so that blockchain supply chain management solutions are already popping up all over the US and abroad as well and more are coming online every day.

Voting: One area that is extremely primed for disruption is the way that voting is handled, not just the voting process itself but everything from the moment a person first registers to vote to every point of contact that they have with the system after the fact. This transition is likely going to happen sooner than later as the 2016 election proved that the current system isn't immune to outside influences no matter how much elected officials wish that was or wasn't the case. Under these circumstances it is clear why a ledger that is viewable by anyone, naturally resistant to malevolent hacking attempts and accessible from anywhere may look so appealing when it comes to making individual elections more representative of the true will of the people and thus make the world a more democratic place as a whole.

There are many different takes on this system currently in the works including Democracy Earth from Buenos Aires which is a company that is working to create a blockchain that will both serve as a means of identification and an online system of voting. In the United States, Follow My Vote is working to create an online voting system that governments can feel comfortable supporting. Besides making the voting process easier, safer and more transparent, it also makes the logistics related to verifying and sorting votes a non-starter as all of that information is automatically handled and sorted by the blockchain. Once smart contracts are added to the scenario the blockchain could even notify winners and losers of the results the minute the polls close.

Chapter 14: The Future of Blockchain 2018

Innovation

Analyst reports from the end of 2017 estimate that the current market worth of the blockchain market is around $410 million. These same reports estimate that the market will grow to a valuation of nearly 10 billion dollars by 2022. This growth is likely going to be driven by a rising demand for blockchain solutions in a wide variety of industries that are looking for improved efficiency and speed in all the processes that blockchain is uniquely suited to improve upon.

Perhaps unsurprisingly, payments and other financial transactions are expected to take up a majority of this space while the largest segment of growth for the market is anticipated to come from Asia. Noted European consulting company Accenture has been doing its own research into 2018 as well and it sees it as the year that blockchain technology moves into the adoption phase. The company went so far as to anticipate that blockchain will be widely adopted by the end of 2018 by major banking organizations around the world.

A similar study was performed by Infosys Finacle who polled leaders from 75 of the world's top financial institutions. They found that about half of these organizations were already investing or looking to invest in blockchain technology. While about 50 percent expected to see some type of mainstream adoption at the commercial adoption by 2020, a full 30 percent expected to have some type of blockchain functionality up and running in 2018.

While interest in cryptocurrency investments is leading to high risk and high reward situations that are good for gaining publicity, there is already an increasing interest in blockchain companies as a stock pick that is high in growth potential while at the same time representing something far more stable than the cryptocurrency alternative. Companies using the technology for things like data management and copyright verification are very likely to see solid returns, taking those who invest along with them.

In fact, research and consulting firm IDC has released its 2018 Worldwide Health Industry Predictions report which states that as much as 20 percent of the health organizations that are currently working on blockchain pilot programs will move beyond that stage to a full rollout before the end of 2020 so 2018 is sure to see some serious movement on that front as well. When it comes to putting promised blockchain's into action, Insurance company AIG also plans on introducing smart contracts into the insurance policy creation system by the end of the year as well. Once one company in each of these spaces is able to start advertising the benefits of their blockchain the floodgates will open, and others are sure to follow.

As such, the research firm Statista predicts nearly $5 billion worth of new investment will reach American fintech companies in 2018 and a large portion of that will be put towards blockchain companies directly. 2018 will also likely mark the year where blockchain technology starts to become associated with more than just cryptocurrency as many of the companies and applications discussed in previous chapters start to come online, more and more people will start to become comfortable with blockchain technology, and as a result usage will continue to increase dramatically.

Also aiding growth will be the continued expansion of cryptocurrency of all forms. While the price of bitcoin continues to tantalize those who failed to get in while the getting was good, a number of other cryptocurrencies are likely going to start seeing more serious consideration from investors as they start to look for options to branch into that are somewhat less high stakes than bitcoin. Other cryptocurrencies to watch include ether, lumens, ripple and litecoin.

Oversight

While it is too soon yet to say what their impact will be, 2018 is also likely to be the year that serious oversight comes to blockchain as well. Perhaps this is for the best, as there are already an ever-increasing number of fraudsters and Ponzi schemes out there that have latched on to the jargon surrounding blockchain to rehash the same old schemes with a new coat of technobabble slathered on top. What's more, the potential for crime inherent in cryptocurrencies is well documented and concerns about this facet of the technology need to be definitively dealt with before mass adoption can take place.

What little regulation that has been put into place so far has all been based around penalties and accidental standards set by companies that suddenly found themselves in positions of market leaderships. 2018 then, presents an opportunity for a new way forward to be established where-in key members of the community are able to work together to develop real, measurable, standards. Ideally, this will lead to a self-regulatory organization, similar to that which the videogame industry developed in the 1990s when concerns first arose of content that was inappropriate for all audiences. Only by working with lawmakers with the blockchain community be able to avoid regulations that do more harm than good.

Regulation: Regulation via enforcement is a popular alternative to lawmaking as the lawmaking process is often exceedingly slow. This is due to the fact that the Administrative Procedure Act requires not only official notice but also a comment period somewhere between the time that a new rule is proposed and when it is officially published in the Federal Register which is required in order for it to be an enforceable law. The fact that each new law requires a comment and notice period as well can easily stretch such things out for a year or more.

This timeline also allows for each of the government agencies which could have something to say about the creation and sale of tokens, including the Financial Crimes Enforcement Network, the Commodity Futures Trading Commission, the Securities and Exchange Commission and more to all have their say before anything is officially accepted. By the time all of them have had their say it will easily be at least 2020. As such, the only way for these regulating agencies to show any control over blockchain technology, in real time, is to announce a new interpretation of existing law and issue serious fines and penalties in hopes that people will get in line without reacting too harshly to the change.

For example, in 2013 FinCen drew the acceptance and transmission of convertible virtual currencies close enough to take bites at the industry. The SEC barked in its DAO report and warned cryptocurrency exchanges not to list security tokens without the proper registration, and recently warned issuers to avoid promises of exchange listing.

Regulation via enforcement is not the same thing as actual lawmaking, however, and is instead akin to ruling via fiat. True rule making on the other hand, typically involves a more careful examination of the issues, detail inquiry into the way the new market is developing, and, most importantly, time, which is why new proposals only come about infrequently. The average report from one of the aforementioned agencies can easily reach 800 pages or more, and can often include pages of comments and suggestions for each subsection of the report.

As this snail's pace is going to be nowhere near fast enough to keep up with the rapid changes that likely await the technology in the new year, it makes 2018 the ideal time for the industry to step up and to start regulating itself. If the opportunity to set their own standards isn't seized enthusiastically with both hands then it is likely that the time will pass for the industry to create its own guidelines, forcing it to take whatever regulations lawmakers decide to give it.

Leadership: 2018 will also likely be the time when a public face is attached to blockchain technology in a more mainstream way as leaders step up and work to organize the marketSo far, cryptocurrency exchanges have reacted well to issues that have arisen in the market, setting new standards in a thoughtful yet responsive way despite not having an interconnected means of doing so.

Other standards have arisen because other companies want to experience the level of success that the major names in the space have already experienced. Unfortunately, this type of standard leads to blind pattern following without the level of thought that will result in truly effective standards being set in the long-term. In order for the industry to mature properly, it is going to need a more thoughtful set of rules to build upon.

The Uniform Law Commission is currently working with engaged members of the blockchain community to develop standards when it comes to cryptocurrency licensing law with the hopes that it will eventually be adopted across the nation. The SAFT project is working on something similar by developing an opensource framework to develop practical legislation for token sales that will exist within the current law framework. Without codified standards blockchain technology will never reach the level of mass acceptance that is required for it to truly become mainstream.

If it's not careful, the blockchain market will find that government regulation will sneak up on it one way or another as a recent IBM study found that an astounding 90 percent of 200 government leaders from 16 companies indicated their countries planned to invest in blockchain technology in one way or another in 2018. These plans for blockchain extend from regulatory compliance, to contract management, asset management and financial transaction management. Likewise, 70 percent of those surveyed believe that blockchain will have a significant effect on their country in the next five years.

Abroad, Russia is ready to commit to blockchain technology in a very real way as Russian authorities recently announced plans to introduce the CryptoRuble in early 2018. According to currently available information, this new cryptocurrency will not be mined in a traditional sense, and will instead be controlled tightly by the state. It is also important to note that why CryptoRubles will have a parity of exchange with traditional rubles, if the owner of the CryptoRubles cannot provide proof of ownership they will be charged an extra 13 percent levy. The same tax will be applied to any earned difference between the price of the purchase of the token and the price of the sale.

While the announcement means that Russia will enter the cryptocurrency world, it is in no way an affirmation or legalization of Bitcoin or any other decentralized cryptocurrency. On the contrary, Putin quite recently called for a complete ban on all cryptocurrencies within Russia.

The statement from Putin seemed apparently to contradict the earlier comments from other ministers who seemed pro-crypto, but only with regulations, as well as Putin's recent meetings with Buterin and others. Now, with the issuance of the CryptoRuble, the apparent contradiction has been made clear.

Chapter 15: The Future of Blockchain – Long-Term Vision

Capital markets: Interest in blockchain technology has already been expanding for several years in this sector of the market, with interest in the field doubling yearly for the past three years. This is largely due to the fact that most of the changes to this sector have come in the form of front office technology, leaving the back and middle office to get along in more or less the same way they have for the past few decades. This, in turn, creates situations where an asset is bought or sold instantaneously before then needing to sit around for several days before the paperwork catches up and makes it official.

The Linux foundation is already hard at work on a fix for this problem and is working to bring together blockchain technology and capital market companies through a standardization of a variation of the blockchain technology that will support the existing capital market infrastructure as much as possible. Ideally, this will result in a scenario where a majority of the remaining inefficiency is removed from the system.

As an added bonus, it will also make it possible for those in the field to offer new, and improved ways of providing services to clients while also allowing regulators to determine new and improved ways to optimize settlement and execution times. This will also come along with an increase in transparency that was previously not only unheard of, but impossible. Much of this will come about as a natural result of the way that smart contracts can work to improve efficacy across all levels of the process.

Banking: When it comes to seeing the potential for blockchain banking in the near future, all that is required is a quick look at China who, in 2017, announced that they were in the process of testing their own form of cryptocurrency in transactions between the People's Bank and other commercial alternatives. While many of the details regarding this new cryptocurrency remain unclear, the information that is available indicates that is likely to be rolled out alongside the renminbi, though a firm timetable is still unknown.

This launch will mark a huge step forward towards the legitimacy of blockchain technology and will truly show that cryptocurrency is on its way to being mainstream. It will also likely do wonders for the renminbi as users will be able to purchase it anywhere in the world and have all the benefits of any other traditional fiat currency while also taking advantage of everything that makes cryptocurrency in its current form so useful. It will also help to serve as an interesting proof of concept for national cryptocurrency as a whole because the challenges that Chinese banks will face will have to eventually be overcome by banks everywhere.

China will also see many unique benefits, starting with the fact that their cryptocurrency will allow economists previously unimagined access to the financial data of the country at an extremely granular level. Just what this level of detailed financial data will reveal is still anyone's guess. Even better, the ease of use with which a cryptocurrency can be used will mean that this will mark the first time that millions of Chinese citizens will have an analog for the types of banking services that much of the world takes for granted.

Digital Transactions: With the prevalence of digital transactions that it is regularly used for increasing each day, blockchain technology is likely soon going to reach a point where the core values that it was released to promote are put to the test. Specifically, the US Federal Reserve is in the midst of designing its own cryptocurrency that is internally being referred to as Fedcoin. The Federal Reserve has already held numerous closed-door meetings with members of the blockchain committee, some of which have been overseen by the chairperson of the Federal reserve herself.

If instituted in the most likely way, Fedcoin will serve to solve the problems that the US government has had with cryptocurrency for the better part of a decade, specifically the fact that it is an obvious outlet for those who are looking to engage in illegal activities online and aren't keen to leave a trace of their activities. This, in turn, means that when it is finally offered to users it will be at the rate of one to one coins to dollars.

Where things start to get complicated, from an ideological level, is that by creating its own cryptocurrency the US government would then have the ability to alter a blockchain once a block has been verified which essentially goes against one of the core tenants of blockchain technology as a whole. Adding this ability to a blockchain will also serve to call its overall legitimacy into question, hurting users' ability to trust it and other similar blockchains as a result, just what the result of this lack of trust will turn out to be remains to be seen.

In general, the Fedcoin blockchain will work the same way as any other blockchain, aside from the major obvious difference. Additionally, it will remove all anonymity from the blockchain, demolishing another long-held tenant of the technology as well. This will also likely have the effect of putting the use of paper money on a ticking clock as Fedcoin will be easier to track than traditional money as well. The public reception to the rollout of this currency will determine a lot about the way that blockchain technology will be used for digital transactions in the future.

Real Estate: Real estate transactions have a well-earned reputation for being extremely tedious and painstaking to undertake, in large part due to the fact that the industry has experienced very little innovation since the advent of the internet. Blockchain technology is well suited to bringing the industry into the twenty-first century, starting with the listing process. With the right smart contract, as soon as a property hits the listing service blockchain it could be automatically sent to those who are searching for a property that meets its qualifications. Once it becomes commonplace it could practically remove property agents and listing services from the equation entirely.

Instead, buyers, sellers, firms and agents will all be able to interact on one blockchain on an even playing field where anyone will be able to both list or complete real estate transaction around the world without worrying about any third party obfuscating the process. Assuming this platform is built on the Ethereum platform, or another blockchain that promotes application use, then it will also allow for a virtually endless number of apps that will be able to take advantage of the properties that such a platform provides.

As an added bonus, getting rid of the traditional centralized structure will free those on all rungs of the real estate profession to experience with a far greater range of fees than those visible on the market today as they will have far more control of the fees that makes sense on a personal level. The only fees that will be associated with listing on the blockchain will be the ones that come along with verification services and keeping the blockchain up and running. At the same time, it will ensure that buyers have easy access to the latest listings without jumping through too many hoops while also making sure that individual sellers have access to the greatest number of interested buyers possible.

Public services: The myriad of individual organizations that make up the public service sector is enough to make the entire industry a labyrinthine mix of rules and regulations that often makes it difficult for those in charge of providing services to actually go ahead and provide those services. This is largely due to the fact that there is frequently no good way for different departments to share their data. This process is often only exacerbated as department budgets are slashed and the way that services are provided or the services themselves are always being shuffled about.

As blockchain technology continues to become more mainstream, it becomes more and more likely that it will be used to address these types of inefficiencies directly. When properly given the chance, blockchain technology will competently serve as an official registry for numerous different types of things that may require a government license to look at officially. It will also come in handy when it comes to coordinating and streamlining the purchasing process for a wide variety of products, ensuring that each government dollar stretches to the absolute limit. Across the board, it is also sure to improve response times while also reducing the risk of fraud and errors, while at the same time improving productivity and efficiency across virtually all levels of the bureaucracy. As a general rule, wherever governmental inefficiency can be found, a blockchain can be used to stamp it out once and for all.

Industry: Modern business tends to run on the backs of those in a wide variety of administrative positions who do little more than manage various databases and ensure that numbers are recorded properly. Auditing firms, auditors, solicitors, supervisory boards, indeed most of the financial sector exists based on the need for third party verification for some type of transaction or another. As such, the biggest disruption that it will cause in this instance is the removal of a need for most of these services completely.

This improved method of verification is going to create change across a wide variety of industries as distributed ledgers offer up a chance to improve the overall level of trust in each system it is connected to. As every transaction is going to be instantly visible to everyone who is a part of it, it naturally ensures that every contract and even every payment is going to be much more trustworthy than its contemporaries that are made through more traditional systems.

This will then likely result in an extreme shift of power away from those who are in charge of keeping an eye on these transactions, though this will likely not benefit new businesses as much as they might expect. Rather, it will be existing business who will be able to leverage their existing resources in new ways that are going to see the most benefit from this new and improved way of doing business. When done correctly, it will help to ensure that they end up in new, and more profitable, positions than they were previously in.

All told, this will serve to increase the overall rate of adoption for blockchain technology as a whole. This is due to the fact that blockchain technology is inherently social in nature. As such, the more people who use it, the more useful it becomes and so on and so forth until the technology reaches a mass saturation point where more than 50 percent of the population interacts with at least one blockchain a day.

Conclusion

Thank you for making it through to the end of this book, let's hope it was informative and able to provide you with all of the tools you need to achieve your goals, whatever it is that they may be. Just because you've finished this book doesn't mean there is nothing left to learn on the topic, expanding your horizons is the only way to find the mastery you seek. Don't forget, blockchain technology as a whole is still in the midst of its very early days which means that new, and potentially game changing, developments could come along at virtually any time. As such, the only way to ensure you don't miss out on any of the good stuff is to become a student of blockchain technology, not just now, but in the long-term.

The next step, then, is to stop reading already and to start considering the many ways that blockchain technology can be put to work in order to change your life for the better. When considering its potential, however, it is still important to not approach the technology with rose-colored glasses as it is not an automatic cure-all for the ills of the modern age which means there will be plenty of instances where it is not the best choice in the moment for what needs to be done. With careful application, however, there is no telling what it might be able to accomplish.

Check out other Books by Matthew Connor

Ethereum - Ultimate Guide

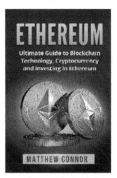

Bitcoin - Ultimate Guide

One Final Thing...

Did You Enjoy and Find This Book Useful?

If you did, please let me know by leaving a review on AMAZON. Reviews lets Amazon knows that I am providing quality material to my readers. Even a few words and rating would go a long way. I would like to thank you in advance for your time.

If you didn't, please shoot me an email at matthewconnor@bmccpublishing.com and let me know what you didn't like. I maybe able to change or update it.

Lastly, if you have any feedback to improve the book, please email me. In this age, this book can be a living book. It can be continuously improved by feedback provided by readers like you.

About The Author

Matthew Connor is a financial technology analyst and a self-taught computer programmer that currently lives in New York City. After graduating from Princeton University with an MS in Computer Science, Matt is currently working for a Fortune 500 company in Manhattan, NY. Matt is passionate about numbers and likes to analyze data to find trends and patterns. Having made his 1st million from investing in Bitcoin, Matt believes cryptocurrencies will revolutionize the world within the next 10 years. Therefore, he is setting out to share what he had learned so others can also get ahead start too. During his spare time, Matt enjoys hiking, reading, and cooking exotic recipes.

Made in the USA
Lexington, KY
10 April 2018